NIGHT NIGHT, SLEEP TIGHT

NIGHT NIGHT, SLEEP TIGHT

Hallie Ephron

wm
WILLIAM MORROW
An Imprint of HarperCollins*Publishers*

This book is a work of fiction. The characters, incidents, and dialogue are drawn from the author's imagination and are not to be construed as real. Any resemblance to actual events or persons, living or dead, is entirely coincidental.

HarperCollins books may be purchased for educational, business, or sales promotional use. For information please e-mail the Special Markets Department at SPsales@harpercollins.com.

FIRST EDITION

Designed by Jamie Lynn Kerner

Library of Congress Cataloging-in-Publication Data has been applied for.

ISBN 978-0-06-211763-2

15 16 17 18 19   OV/RRD   10 9 8 7 6 5 4 3 2 1

For Molly, Naomi, and Frances Louise

ACKNOWLEDGMENTS

One of the pleasures of writing this book was taking a trip back in time, remembering what it was like to grow up in Beverly Hills. A special thank-you to the folks in the "Beverly Hills in the 50s & 60s" Facebook group for sharing their remembrances, as well as to friends Jodyne Roseman and Ellen Kozak. For a sanity check on Hollywood and the movie business, thanks to my sister Delia Ephron.

Thanks to Lee Lofland for help on police procedure; to Deb Duncan on insurance fraud; to Paula Shelby on the workings of a Harley dealership; to Susannah Charleson on arson investigation; to Clarissa Johnston, MD, on trauma; and to Michelle Clark on death investigation.

For help working my way out of plot holes, thanks to generous fellow writers Paula Munier, Roberta Isleib, Hank Phillippi Ryan, and Jan Brogan.

I am deeply indebted to my agent, Gail Hochman, and my editor, Katherine Nintzel, for their clearheaded critiques and encouragement. Seriously. Thank you. Thanks to assistant editor Margaux Weisman for help shepherding this manuscript through to publication.

Thanks to Joanne Minutillo, Danielle Bartlett, Tavia Kowalchuk, and the other amazing folks in publicity at HarperCollins for their talent and enthusiasm launching this book.

Thanks to Jim and Anne Hutchinson, whose generous donation to Raising a Reader purchased a character name for their son Jack.

And finally, thanks to my husband, Jerry. Without his patience and forsaken weekend outings, this book never could have been written. He, more than anyone, was glad to see it finished.

NIGHT NIGHT, SLEEP TIGHT

FRIDAY,
MAY 23, 1985

CHAPTER 1

Arthur Unger slides open the glass door and steps out onto his flag-stone patio. He's had a few drinks but he doesn't feel them. It's late at night, and though the sky is clear and there is no moon, there are no stars, either. There never are. Between ambient light and air pollution, he'd have to drive to Mount Baldy to see Orion's Belt.

The sky is . . . He gazes up at it. *Opaque? Inky? Like warm tar?* His ex-wife would have nailed it. She was great at narrative description and dialogue. And of course, she could type. He was the plotmeister. Arthur takes a final drag on his cigarette, the tip glowing in the dark, and stubs it out in one of the dirt-filled, terra-cotta planters in which Gloria once cultivated gladiolus. Or was it gardenias? Something with a G.

He picks up one of three cut-glass tumblers sitting on the table on the patio, left over from tonight's unpleasantness. Why does he have to rehash what was agreed on and settled years ago? He did what he promised. DEBT PAID IN FULL should be stamped across his forehead, and he has the paperwork to prove it.

He raises the glass in a toast. *To the end of old debts and life without gardenias.* He knocks back lukewarm, scotch-flavored ice melt, then reaches into the house and flips a light switch. The water in the pool—it's just twenty-five meters long, not the size to which he feels entitled, not what he expected to have earned by this point in his life—glows radiant turquoise against a row of coral tiles above the waterline.

Arthur imagines his yard is a movie set. A camera dolly backs up in front of him as he strides across the lawn in terrycloth slippers, an open Hawaiian shirt, and bathing trunks. A pair of amber-tinted swim goggles hangs loose around his neck. He reaches to unlatch the gate in the utilitarian chain-link fence surrounding the pool, but it's already open. Careless. He once had to scoop a neighbor's Chihuahua from that pool, and he has no desire to fish another dead animal from the water.

He slips through the gate and latches it behind him. Tosses a towel over a chaise longue made of aluminum tubing and white vinyl strapping. Takes off his shirt and drapes it there, too.

Arthur is in his late fifties, conscious of once taut muscles in his chest that sag if he exhales and relaxes at the same time. Even alone in his own backyard he tries not to let down his guard. Tonight he looks tired, the dark pouches under his eyes echoing the flab in his gut. He needs his thirty laps to drain away anxiety, get his blood flowing, and make him feel sufficiently worn out to fall asleep without a Valium and another scotch.

The pool has been skimmed, at least. That's supposed to be his son's job, but Henry rarely notices that it needs skimming. Rarely notices much at all, in fact. Henry seemed stunned when Arthur told him the house has to go on the market, though this would be patently obvious to anyone living here and even minimally aware of anyone's needs but his own.

Time to grow up, baby boy.

That's a line from a musical comedy Arthur and his wife wrote. *Show Off* was supposed to star Judy Holliday and Dean Martin, but she dropped out to have throat surgery. And then, of course, there was the breast cancer. Tragic for her. Tragic for audiences to lose such a brilliant comedic talent.

Arthur first met Judy way back when . . . He closes his eyes and tries to remember. He must have been working as assistant stage manager at the long-ago razed Center Theatre in Manhattan, a gorgeous art deco ark at Forty-Ninth and Sixth. Late nights, after the show, he'd take the subway down to the Village where Judy performed a cabaret act with Betty Comden and Adolph Green, accompanied at the piano by fellow unknown Leonard Bernstein. They were all so young. So talented.

Show Off could have been big. Should have been big. Would have kicked his and Gloria's career back into high gear. Plus he'd have scored a producing credit and points on the back end. Even with the studio's creative bookkeeping, eventually it might have earned him a decent-sized pool and enough in the bank that he could have offered to help out his kids when they needed it.

But he's not out of the game yet. His book could catapult him back onto the A-list. It's the quintessential Horatio Alger story. He'll write the screenplay. Direct the film version. Cast a great actor in the lead. Someone capable of nuance. Subtlety. A little comedic flair. Maybe he'll give himself a walk-on cameo.

Arthur runs his fingers back and forth through his hair, still dark and thick and curly, about the only good thing that both he and Henry inherited from Arthur's father. Then he stretches, arms wide, fingers splayed. Inhales deeply. Coughs. He tries to imagine his girls, *ingenues* as they were once called, perched around the edge of the pool. He sees the camera panning from one to the next, sliding appreciatively across cleavage and shapely leg, then over to him as he smiles benevolently back at them.

Give a little, get a little. That's always been his motto. Only lately he's been getting just that. A little. And sometimes not even that.

He adjusts his swim goggles over his eyes. His girls, if they were really there, would be a golden blur now. The camera dolly would have faded into murky darkness. He imagines it tracking him as he steps slowly, deliberately to the deep end of the pool. The wall of his garage, lit by a wavery glow, makes an eerie backdrop as his shadow creeps up it, nearly reaching the windows of his second-floor office. In a film, it would feel like foreshadowing. Very dramatic. Perhaps a bit melodramatic—at least that's what Gloria would have said, and she'd have been right.

Arthur faces the pool. Centers himself. Three long strides and he feels the concrete apron around the pool under his bare foot. A leap and he's airborne, outstretched and arcing in a racing dive. He lands more heavily than he'd like and the water is a lot colder than he expects, but the initial shock quickly recedes.

He swims, stroke after stroke after stroke, a turn of his head to take a breath. He reaches the far end, barely pausing before pushing off and surging in the opposite direction. Back and forth. He's nearing the zone, the place where his mind lets go and muscle memory takes over. He starts to review his work in progress the way he used to work through a movie script, running the maze of major and intermediate plot points; unpacking the emotions, goals, and obstacles that drive them; probing at knots and dead ends until he'd worked them through.

But what rises to the surface, like the taste of a bad oyster, is last week's phone call and tonight's meeting. Arthur feels his shoulders tensing up, his breathing begin to labor. If he wanted to be dictated to, told what he could and couldn't write, he'd be working on a screenplay.

He's not looking forward to tomorrow's meeting with the Realtor. She advised him to list the house at 899K. Apparently nine hundred is a barrier for buyers, and even though the market is hot, his house is not. How did she put it in the ad copy? *A classic open-plan home with lots of possibilities.* In other words, a dump. But hey, it's Beverly Hills, even if it is in the flats south of Sunset, and Arthur is determined to list his house for an even mil. Let them underbid and think they're getting a bargain. One thing he's learned: you don't ask, you don't get.

His daughter, Deirdre, agreed to drive up from San Diego tomorrow and spend a few days helping him get the place ready to go on the market. He called to remind her before coming out for his swim, but she didn't pick up the phone. Probably sleeping—Arthur loses track of the time. It's no big deal because Deirdre has never needed reminding. Even when she was little, she got out of bed and dressed for school without having to be coaxed. She did her homework. As grueling as the physical therapy was, she just did it, never complaining, even when it was clear that it wasn't going to make any difference.

Once he'd have taken Deirdre for granted. But having someone he can depend on is something he cherishes now. Especially since he hasn't exactly earned her undying loyalty. She blames him for the car accident that crippled her, and how could she not? It's the kind of thing that apologies can't fix, though Lord knows he's offered them up.

Apologies. Excuses. Anything but the truth. It really wasn't his fault, but he can't tell her that, because if it wasn't his fault, then whose was it? He can't go there. Not yet, anyway. Maybe never. Besides, it's the last thing Deirdre wants to hear and it won't bring her the peace he wishes he had to give her.

He strokes across the pool. Turns.

He means to thank Deirdre properly this time. Maybe send flowers to that art gallery of hers. Her constancy is such a support these days. Henry seems incapable of fulfilling even the smallest commitment unless it involves one of his dogs or his muscle bikes. Gloria left years ago. Oh sure, they talk on the phone, though only occasionally, and not since Gloria began a monastic retreat. Tibetan Buddhist, shaved head, vegan diet, the works.

He tries to let go, to push away unfinished business as he pushes off from the end of the pool. As he strokes, he tells himself, *There's always tomorrow*. A wrong-answer buzzer goes off in his head. As Gloria often chided him: *Focus on the now*.

In this moment, as he swims in a steady rhythm, he is the star of his own movie, his life story brought to the silver screen, his backyard the set. The director, hidden in darkness behind the camera, has long ago called for *Quiet! Roll camera! Action!*

Fade in.

All attention is focused on Arthur as he turns again at the far end of the pool.

A beat.

He plunges back in the opposite direction, stroking powerfully toward the spotlight where . . . Who? Billy Wilder and Elizabeth Taylor, he decides, are waiting for him to surface and accept a golden statuette with his name engraved on the base. Recognition of his lifetime achievement, something even his kids can feel proud of.

Ready for his close-up, Arthur reaches the opposite end of the pool, raises his head, and hangs there for what he thinks will be just a moment, basking in the illusion that he's the star of his own show. Realizing too late that the spotlight he sees through the goggles—a yellow, water-streaked glow—is real and getting bigger, until it's bright and blinding and right in his face.

He blinks and looks away, and in that split second the light goes out. And we hear the sound that Arthur can't—the thud of heavy metal connecting with Arthur's head, his prefrontal cortex to be precise, the part of his brain responsible for a lifetime's worth of lousy decisions and selfish moves.

It's a wrap.

SATURDAY

MAY 24, 1985

CHAPTER 2

By the time Deirdre Unger reached the Sunset Boulevard exit off the San Diego Freeway, her stomach burned. The Egg McMuffin she'd wolfed down an hour and a half ago had been a mistake.

Used to be this was an easy turn, but traffic had grown heavier over the years. As she waited, she took a sip of what was left of her coffee. It tasted mostly of waxy cardboard and only made her stomach seethe. She set the cup back in the drink holder and foraged with one hand in her messenger bag, feeling for an errant Rolaid or Life Saver and coming up with only lint.

"How hot is it, kiddies?" The voice on the radio sounded maniacally overjoyed. "So hot trees are whistling for dogs!" A buzzer sounded, then hollow laughter. "Seriously, it's hot out there, so drink plenty of water. Red flag warnings have been issued for today and tomorrow. Heat and dry winds are expected to turn Los Angeles and Ventura County mountains and valleys into a tinderbox."

Yippee. Deirdre snapped the radio off and gripped the wheel. Another reason to have stayed in San Diego.

At last there was a break in the traffic and she turned onto Sunset. Why on earth was she doing this? Couldn't Henry for once in his life have stepped up to the plate? She wondered, what would he do after the house sold? No way he'd want to live with Arthur in a condo complex filled with actual grown-ups. He'd have to find a place for himself and Baby and Bear—those were his rottweilers—and his Harleys. She had no idea how many bikes he had at the moment, but she wouldn't have been at all surprised if he'd named them, too.

It was a shame about Henry. He'd wanted desperately to be a jazz guitarist, and if he'd worked at it, he might even have made a career of it. But freshman year of college he dropped out, stopped playing, and moved home. Not that he'd done badly after that. He made a good enough living selling bikes for a Harley dealership in Marina del Rey. Problem was, he "invested" his earnings in vintage bikes, Stratocasters, and the best pot that money could buy. Girl-friends came and went so fast Deirdre had stopped asking. Henry seemed to be allergic to any kind of personal commitment.

A loud *blat* came from a passing car. Deirdre realized she'd nearly sideswiped it. She jerked her car back in its lane. *Get a grip,* she told herself. Her father had asked for her help. He'd mellowed a lot in his old age, and even took the occasional break from his monologue to ask what she was up to. And it was just a weekend, not a lifetime.

She'd intended to drive up last night, but at the last minute her business partner, Stefan Markovic, got a call from an arts reporter for the *Wall Street Journal* who wanted to meet with him to talk about the new arts district that was taking shape in San Diego. She and Stefan had agreed it was potentially great publicity. But that meant he wasn't there to help install their new show, so she'd been at the gallery with the artist's assistant until after midnight. By then it was too late to start driving to L.A., so Deirdre had gone home. Before she went to sleep she'd turned off her phone's ringer. Her father had a nasty habit of calling at all hours of the night, using her silence as permission to rattle on about his latest brilliant idea or vent his spleen, depending on how much he'd had to drink. When he was done, he rarely said good-bye. He'd just hang up, and she'd end up lying in bed for an hour, trying to fall back asleep.

Deirdre crossed into the left lane and accelerated. Power surged and her Mercedes SL automatically downshifted and shot forward, hugging the road as she pushed it around a bend. She braked into the curves and accelerated coming out, weaving between cars on the winding four-lane road. Forty, forty-five, fifty. The end of her crutch slid across the passenger seat, the cuff banging against the door.

The car drifted into the right lane coming around a tight curve and she had to slam on the brakes behind a red bus that straddled both lanes and poked along at twenty miles an hour, idling just outside walled estates. STARLINE TOURS was painted in slanting white script across the back.

Deirdre tapped the horn and crept along behind the bus, past pink stucco walls that surrounded the estate where Jayne Mansfield had supposedly once lived. It had been a big deal when the actress died, had to have been almost twenty years ago. And still tourists lined up to gawp at her wall. Breasts the size of watermelons and death in a grisly car accident (early news reports spawned the myth that she'd been decapitated)—those were achievements that merited lasting celebrity in Hollywood. That, or kill someone. It was the same old, same old, real talent ripening into stardom and then festering into notoriety. Deirdre sympathized with Jayne Mansfield's children, though, who must have gone through their lives enduring the ghoulish curiosity of strangers.

Buses like the one belching exhaust in front of her now used to pull up in front of her own parents' house, passengers glued to the windows. Most writers, unless they married Jayne Mansfield, did not merit stars on celebrity road maps. And in the flats between Sunset and Santa Monica where her father lived, notables were TV (not movie) actors, writers (not producers), and agents, all tucked in like plump raisins among the *nouveau riche* noncelebrity types who'd moved to Beverly Hills, so they'd say, because of the public schools. You had to live north of Sunset to score neighbors like Katharine Hepburn or Gregory Peck. Move up even farther, into the canyons to an ultramodern, super-expensive home to find neighbors like Frank Sinatra and Fred Astaire.

Arthur Unger had earned his spot on the celebrity bus tour through an act of bravery that had lasted all of thirty seconds. It had been at a poolside party to celebrate the end of filming of *Dark Waters*, an action-packed saga with a plot recycled from an early Errol Flynn movie. Fox Pearson, the up-and-coming actor featured in the film, either jumped, fell, or was pushed into the pool. Sadly for him, no one noticed as the cast on the broken leg he'd suffered a week earlier doing his own stunts in the movie's finale dragged him to the bottom of the deep end. Might as well have gone in with his foot stuck in a bucket of concrete.

A paparazzo had been on hand to immortalize Arthur shucking his shoes and jacket and diving in. Fox Pearson's final stunt, along with its fortuitous synchronicity with the movie's title, earned more headlines for the dead actor than any of his roles. Suddenly he was the second coming (and going) of James Dean, a talent that blazed bright and then . . . cue slow drumroll against a setting sun . . . sank below a watery horizon.

When talking about it in private, Arthur liked to quote a line from *Sunset Boulevard*. "The poor dope—he always wanted a pool. Well, in the end, he got himself a pool."

Deirdre used to dress up in her mother's silver fox stole and wave at the bus from the window seat of their dining room. She perfected an open handed, tilt-to-tilt wave like one of those gowned-up girls in the Rose Parade. Back then she could dream of being in the royal court. Queen, even. But beauty queens didn't have withered legs.

Finally the bus pulled over so that Deirdre and all the cars backed up behind her could pass. A few minutes later she cruised past the familiar brown shield, its message printed out in gold letters: WELCOME TO BEVERLY HILLS. After that, the twisty road straightened into a divided parkway and the speed limit dropped to thirty, as if chastened by the wealth surrounding it. There was not a single pedestrian on the sidewalks. Not a soul in the crosswalks or waiting at bus shelters.

A half-dozen blocks farther along Deirdre turned south. Two blocks down, she pulled over and parked in front of the house where she'd grown up: stucco façade, front courtyard, and arched living room window screened by an elaborate wrought-iron grille. That was Henry's black Firebird parked in the driveway. Arthur kept his red TR8 in the garage. To the casual observer the house seemed the same as it had for years. Decades, even. She could imagine the ad: *Charming one-story Spanish colonial, three bedrooms, two and a half baths, in-ground pool.*

Deirdre sat there for a few moments, listening to the car's engine tick in the silence and wishing she wasn't such a compliant daughter. Then she reached for her messenger bag, looped the strap over her head and across her chest, and grabbed her crutch. She climbed out of the car and leaned against the door. Heat seemed to pulse off the macadam. She put on her sunglasses and took a harder look at the house. Terra-cotta roof tiles were missing, and the once white exterior was more the color of weak tea. Deirdre doubted it had been painted since her mother left, the last time for good, nearly twenty years ago. Maybe that real estate ad should include the chipper warning: *Fixer-upper.*

Not that everyone fixed up Beverly Hills houses these days. Parcels of land had become so much more valuable than the houses on them, why bother? Buyers tore down and started over, erecting new houses that looked like they were worth the million or more you had to shell out to get an address with a 90210 zip.

Case in point: Across the street from her father's house, where there had once been a gracious, one-story Spanish colonial, there now sprawled a house worthy of a southern plantation. Two-story columns and Palladian windows flanked a magnificent pair of coffered front doors: Tara with vertical blinds, and badly out of scale for its third-of-an-acre lot.

Several more properties on either side of the street had been similarly perverted, and another was in process. Her father's house, once typical for the neighborhood, had turned into an anomaly.

Deirdre popped the trunk and slammed the car door. She eased her arm through the crutch's cuff and grasped the grip to which she'd duct-taped an extra layer of foam padding. She stumped to the back of the car and pulled a small duffel bag from the trunk. She'd packed light.

As she crossed the lawn she felt the rubber tip of her crutch sink into the grass. It made a little popping sound as she pulled it out. The courtyard was a tad cooler, shaded by a leaning olive tree. The ground under it was awash in rotting olives, some of them squashed and bleeding red slime on the gray stepping-stones. Deirdre knew from experience they could be treacherous to her crutch, so she picked her way carefully around them.

Many of the blossoms on the pair of camellia trees, one planted when Henry had been born and another about a year later for Deirdre, had turned brown and rotten, their season ended, though Deirdre's tree still bore white camellias. Once smaller than she was, the tree was now about ten feet tall. It was probably the only thing she wanted to take when the house was sold. She hoped it could survive being dug up and transplanted in the backyard of her little bungalow in Imperial Beach.

Deirdre tried the front door. It was locked, so she had to ring, which set off Henry's dogs. She didn't have a key to the house because Arthur kept forgetting to send her a set. That was his way, everything always and forever at his convenience.

When still no one answered the door, Deirdre knocked again, then rang some more. The dogs were going bananas. None of it roused anyone. Now what?

She dropped her duffel on the front step and walked back across the courtyard, trying not to slip on the olives or get the tip of her crutch stuck in the pillowy moss that grew between the stones. On the driveway the air was fifteen degrees hotter. A shovel was lying behind Henry's car. Deirdre picked it up, leaned it against the two-car garage, and peered in through one of the little windows in the overhead door. Motorcycles, at least two of them, were lined up in one bay. Her father's car was in the other. Which meant he had to be there, too. He was probably in his office up on the second floor of the garage.

Deirdre tried the overhead doors. They were both locked. Then she tried the regular door that led to the stairway. It was locked, too. She knocked. Hollered. Whistled. Was he asleep? She ought to just go over and bang on Arthur's bedroom window. It was nearly noon, for heaven's sake.

She was crossing the yard when she noticed the gate to the pool was open—wide enough for a pet or a child to easily slip through and fall in. Keeping that gate secured was one of the few things that her parents had agreed upon. She was about to go over and shut it when the dogs started up again. There they were, on the other side of the living room's sliding doors to the patio, their claws scratching the glass.

Deirdre went over to them. "Hi there, knuckleheads," she said. Bear whined and wagged his butt where there was the stump of a tail. Baby, who was a little smaller and had a bit more golden brown over her eyes and around her muzzle, woofed and stood up, her front paws resting against glass smeared with doggie saliva. She was nearly as tall as Deirdre.

Deirdre tried to slide open the door, but of course it was locked too. "Dad! Henry!" she shouted. "Would one of you please get out here and open a damned door so I can come in, preferably before one of the dogs has a heart attack. Come on! It's hot as hell out here."

She waited. Someone had been out there not all that long ago: on the patio table sat a cut-glass tumbler with a bit of pale amber liquid at the bottom of it.

The only vestiges of Gloria, who'd long ago walked out on Arthur, were barren terra-cotta pots surrounding the patio. Once they had contained her collection of scented geraniums. Now they held only dried-out soil and the skeletal remains of weeds.

How her mother used to fuss over her prized *specimens*, as she called them, picking off dead leaves and pruning the branches into striking, bonsai-like shapes. Now she grew herbs and taught serenity and was well along on "the path," as she termed it, in the midst of a Buddhist retreat that required her to shave her head and—something Deirdre could barely imagine—remain silent. Deirdre had known her parents' marriage was over when her mother started carrying *malas*, prayer beads, that she fingered in quiet moments as she meditated and whispered mantras under her breath. When she'd moved to the desert commune near Twentynine Palms, she'd taken only one plant with her, a rare hybrid that smelled like smoked chili pepper, abandoning the rest to Arthur's inevitable neglect.

Deirdre turned back to the pool. Her mother had detested that pool and the chain-link fence that surrounded it. She'd tortured Arthur with plans for turning the entire backyard into a Japanese-style garden of raked stones and koi ponds. He'd wanted a sauna and hot tub. It made Deirdre wonder: If it hadn't been for their success as a screenwriting team, would her parents have stayed together even long enough to have had Deirdre?

That's when Deirdre noticed Arthur's favorite Hawaiian shirt draped over the chaise longue by the pool and his slippers on the ground beside it. She couldn't remember him ever swimming laps in the morning.

Behind her, the dogs quieted. She turned back. Henry was there on the other side of the glass, bare-chested and wearing a pair of drawstring sweats that rode low on his hips. The thick gold chain he wore around his neck reminded Deirdre of the choke chains he used to train the dogs. He yawned and rubbed his grizzled face, then unlatched and slid open the door.

The dogs burst from the house and ran joyous victory laps around the yard. Bear leaped for the knot at the end of a rope Henry had tied to a tree branch and hung there wriggling and snarling. Baby circled back to Deirdre, who crouched and let Baby lick her face. She buried her face in the soft ruff around the dog's neck. Whatever else you could say about Henry, he raised the sweetest dogs.

"Yo, Deeds," Henry said, offering his hand and helping her up. He gave her an awkward hug, then stood back and yawned, exhaling stale beer breath. "What are you doing here?"

Deirdre forced a smile. She knew it wasn't fair—after all, how could he have known she was coming if she or Arthur hadn't told him—but the question annoyed her. "Dad asked me to come up and help him with the house." She couldn't resist adding, "He's selling it, you know."

"Yuh." Henry crossed his arms. "I know. Whyn't you ring the front?"

"I rang. I knocked. *Whyn't* you answer?"

"I was sleeping. And besides, Dad's here. Why didn't he—" He turned and bellowed into the house. "Yo, Dad! Where the hell are you?"

Deirdre listened with him, but when the house remained silent, Henry said, "Well, I thought he was here." He shuffled off in the direction of the bedrooms, only to reappear moments later. "So . . . where is he?" He stepped out onto the patio. "Is his car here?"

"Parked in the garage."

"Maybe he's up in his office."

"It's locked. I knocked. And yelled. Looks like he took a swim. He left some stuff out here." Deirdre pointed to the shirt and slippers.

She edged a few steps closer to the pool, then stopped. Her neck tingled and she smelled blood in her nose as she realized that there was a shadowy shape submerged under the water at the deep end of the pool.

CHAPTER 3

Deirdre felt as if, for a moment, the iris of a camera closed and opened again in front of her. *Click.* She dropped her messenger bag and stumbled across the patio, onto the grass, cursing the crutch that made a lousy substitute for a good leg. She was a strong swimmer if she could ever get to the damned pool.

Henry flew past her. In seconds he was across the yard, through the gate, and diving in. He took two strokes underwater and then surfaced, driving the body that Deirdre knew was her father to the side of the pool.

Deirdre reached the edge and sank to her knees. "Oh my God. Daddy?"

Henry held on to the tile edge of the overflow channel, gasping and trying to lift what was surely dead weight. Deirdre grabbed her father under an arm. Between her pulling and Henry pushing they managed to lift him out onto the concrete apron.

Time seemed to slow down as Deirdre shivered and backed away, then sank into a crouch. Her father lay on his side, his back curled and knees bent, hands stiff in front of him as if the water had returned him to the womb. His eyes were open, their surfaces clouded over, and the skin on his hands had shriveled like loose latex. She knew CPR, but anyone could see that her father was well beyond help.

Bear licked her hand. Beside him, Baby was down on her haunches like a sphinx, coat glistening, her massive head tilted, staring at Arthur. Henry was ashen, holding on to the edge of the pool. His lips moved, and she knew that he was saying something, but it felt as if rushing water filled her head.

He was dead. Her father was dead. If only she'd gotten there sooner. If only she hadn't stopped for that Egg McMuffin. They had to call an ambulance. Or the police. Or the fire—

"Deirdre!" Henry's voice penetrated. "Are you okay?"

Deirdre tried to speak, but the breakfast sandwich was backing up in her throat. She burped and her mouth filled with acidic coffee.

"Stay here. I'll call 911," Henry said and hoisted himself out of the pool.

"I'll go," she said, reaching for her crutch.

Her weak leg was folded under her. She struggled to her feet, threw Henry the towel from the lounge chair, and clumped as fast as she could, hand over her mouth, through the gate, across the yard, and into the house. She reached the bathroom just in time.

Afterward, she stood at the sink, splashing water on her face and then drinking from her cupped hands, trying to wash away the nasty aftertaste. She looked into the mirror. Her long dark hair was wild around her face, like Medusa's snakes in the Caravaggio portrait, her father's haunted eyes staring back at her. She shivered, realizing that her dark leggings and top—an oversized T-shirt with XENO ART, the name of her gallery, silk-screened across the front, the neck artfully torn out—were completely soaked.

All she could find to dry her face was a ragged hand towel. She blew her nose and grabbed a few extra tissues for later, tucking them into the waistband of her leggings.

Numb, moving like a defective robot, she limped into the kitchen. The phone hung on the kitchen wall. She punched 911 and sank into a chair at the kitchen table, trying to collect herself.

An operator picked up. "Beverly Hills 911. Where is your emergency?"

Where? Deirdre wasn't expecting the question. It took her a moment to come up with the address of the house where she had grown up.

"Thank you. What's the emergency?"

"My father. He drowned in the pool. He's dead." Her voice sounded as if it were coming from someone else's throat.

"Are you sure he's dead?"

Deirdre closed her eyes. She could see Arthur's stiff, clawed hands. "He's dead."

"Is anyone there with you?"

Deirdre squeezed the receiver. "Please send someone."

"They're on their way. Are you alone?"

"My brother . . . He's—" She stood and gazed through the window, past white ruffled café curtains that she'd helped her mother hang. Her vision blurred. She had to call her mother.

"Hello?" The dispatcher's voice sounded far away. Deirdre was trying to remember where she'd written the phone number her mother had given her months ago, before she'd checked into that Buddhist retreat. It had to have been in her datebook. Which was . . . she tried to recall where.

She hung up the phone, belatedly registering the dispatcher's "Please stay on the line . . ."

She must have dropped her bag on the patio. She went to look. Sure enough, there it was. She opened the sliding door and shouted to Henry, "They're on the way. I'm calling Mom."

A minute later she was dialing, even though she knew no one

would answer—it was a *silent* retreat for God's sake. At least she got an answering machine. "Cho Bo Zen Buddhist Temple. Please leave a message. *Gassho.*"

After the beep, Deirdre said, "This message is for Gloria Unger. I'm her daughter. Please tell her—" What? That something had happened and to please call back at once? No. Her mother would worry that something had happened to Deirdre or Henry. So she just said it: "—Arthur died. Suddenly. He . . ." Deirdre pulled the handset away from her face and stared at it, then put the receiver back to her mouth. "Mom?" Her eyes misted over and her throat ached. "Daddy drowned."

Deirdre ended her message with "It's Saturday," because who knew if there was a date stamp on the phone messages or how often the monks checked the machine. "I'm staying at the house. Henry's here. And I wish you were here, too."

The doorbell chimed just as she managed to croak out, "Please, call back." *Could the police have gotten here that fast?*

Deirdre hung up the phone, wiped her eyes, blew her nose, and went to answer the door. She expected to find paramedics or grim-faced police officers outside. Instead, a woman about her own age stood there, arranging a white bow at the neck of the blouse she had on under the jacket of a dark pantsuit that was a size too small.

"I'm here to see Arthur Unger," the woman said. Her gaze traveled to the crutch Deirdre was leaning against. Deirdre was used to that.

Was that a siren in the distance? Deirdre looked past the woman.

"Deirdre?" The woman wiped away beads of sweat that had formed on her upper lip. She seemed vaguely familiar. Maybe an actress? Arthur was always having hopeful young women over to the house to *read lines*, even when everyone knew that Arthur's only lines were the ones he used to convince the world that he was still a player.

"You don't remember me, do you?" the woman said.

Finally Deirdre really looked at her. Auburn hair. Sloping eyes.

Pale soft flesh and freckles like sugar sprinkled across her nose. Deirdre did remember her. Of course she did.

"Joelen?" *Joelen Nichol.* Deirdre hadn't seen or spoken to her in what, at least twenty years? Not since high school. Not since that night. She was the daughter of glamorous Elenor "Bunny" Nichol, a movie star known for her spectacular silhouette, electric blue eyes, and lousy taste in men. Joelen had confided to Deirdre that her father's name was Joe. That explained her unusual name, pronounced *Joe-Ellen*—a combination of Joe and Elenor.

Joelen had her mother's incredible aquamarine eyes, luminous complexion, and radiant smile with dimples on either side. "It's so good to see you." She grasped Deirdre's arm, oblivious to the sirens that were growing louder. "This is so amazing. I had no idea you'd be here, too. Did he tell you that we had an appointment?"

"He?"

"Your father. I have a meeting with him this—"

"No." The word came out louder than Deirdre intended. Joelen recoiled. "I'm sorry. He . . . he can't see you now. It's too late. He's—" Deirdre couldn't finish it.

"What? Did he change his mind? Is this a bad time?" Joelen started to back away, tripping over her own feet. The siren was screaming now. "I can come back. No problem. Another time?" She pulled a card from the outer pocket of her briefcase, lunged forward, and gave it to Deirdre. "Tell him to give me a call and—" Joelen broke off midsentence when the sirens fell silent. She turned and stared out toward the street.

Deirdre brushed past her. She moved through the courtyard, jerking her crutch loose when it got stuck between the paving stones. A police cruiser was parked in front of the house, lights flashing. Pulled up behind it was a red truck with gold lettering on the side: BEVERLY HILLS FIRE PARAMEDICS.

Deirdre was dimly aware of Joelen scurrying from the house, crossing the street, and getting into a dark compact car as Deirdre pointed two paramedics to the backyard. One of them carried an oxygen tank. Another maneuvered a wheeled stretcher that clattered up the driveway. A pair of uniformed police officers raced around ahead of them. Deirdre trailed behind. Henry was waiting at the gate to the pool. He'd wrapped up in the towel and, in spite of the heat, was shivering. He had the dogs on tight leashes, sitting tensed at his side.

The EMTs raced for the pool. Henry watched them for a moment, then led the dogs back into the house. Deirdre waited on the patio for him to come back out. She crossed her arms, feeling stiff and chilled as she watched one of the paramedics kneel beside her father. The oxygen tank lay abandoned on the ground.

A police officer stood by the pool, talking on a radio. All Deirdre could hear were bursts of static. The officer belted the receiver, exchanged a few words with one of the paramedics. He crouched by the body, then lingered there a few moments longer, looking into the pool.

Slowly, he got to his feet and took in the yard and the back of the house, then shifted his gaze over at Deirdre and Henry. He crossed the grass to the patio. He was an older man with the boyish intensity and short sturdy stature of Richard Dreyfuss.

"I'm Officer Ken Millman." He offered Deirdre his I.D., just like in the movies, only this wasn't a movie. "I'm sorry. He's gone."

Deirdre knew full well that her father was dead, and yet she felt as if the air had been sucked out of her. She groped for a chair and sat. Tears filled her eyes, her stomach clenched, and her mouth opened in a silent scream.

CHAPTER 4

Deirdre barely heard Henry's "Are you okay?" Or the police officer's "Do you need a glass of water?" She tried to say *I'm fine*, to wave them away, but it was another few minutes before she could even lift her head. She found the tissues she'd stashed in her waistband and wiped her eyes. Blew her nose. Sat some more, just trying to wrap her head around what had happened.

At last, she found her voice. "Sorry."

The police officer whose name she'd already forgotten was crouched in front of her, his eyes searching hers. "The county coroner will be here soon." He was speaking slowly. "It's routine in an unattended death. Do you understand what I'm saying?" He waited for her nod, then continued, "I need to collect some information. Are you okay with that? Can you answer a few questions?"

Deirdre blinked. Henry put his hand on her back.

The officer stood. He pulled out a notebook and flipped it open, then thumbed to a fresh page and jotted a few notes. "The victim's name?"

"Arthur Unger," Henry said. "He's our dad."

"He lives here?"

"Yes. I live here, too. Deirdre lives in San Diego." Henry gave the officer his name and phone number. Deirdre gave him hers.

"Can you tell me what happened?"

"We don't know what happened," Henry said. "My sister got here and found him floating in the water."

Not really floating, Deirdre thought. Arthur had been barely suspended above the bottom of the pool, beneath the surface, like a fly in amber. She choked at the memory.

"How long ago?" the officer asked.

"Not long," Henry said. "Fifteen, twenty minutes maybe."

"I called it in right away," Deirdre said.

The officer drew a rectangle on his pad, and around it a larger dashed rectangle with a gap that Deirdre realized was meant to represent the chain-link fence. "Can you show me approximately where your father was when you found him?"

Her hand trembling, Deirdre pointed to a spot near the edge in the deep end.

The officer drew an X. "Then what?"

"My sister called 911."

"You're both wet." The officer squinted at Henry, then looked over at Deirdre. "I'm guessing one of you went in after him."

"Of course," Henry said, looking more annoyed than chastened. "I did. I thought . . . Actually, I didn't think. I mean, it was just a gut reaction. He might have had a heart attack or a stroke. Or fallen in and hit his head, for all we knew."

"I see." The officer gave Henry a long look. Deirdre had the distinct impression that he didn't think Arthur had just fallen in. "Thank you. That's all for now. I need you both to wait in the house until we finish up our investigation out here."

Henry hesitated a moment, then turned and started for the house.

27

"And I need you to leave that where you found it," the officer said, indicating the tumbler that Henry had picked up from the table.

Deirdre sat at the dining table, watching the activity through the sliding glass doors. Investigators had constructed a makeshift tent over Arthur's body. To protect him from what, she wondered. A photographer took pictures—not just of Arthur but of the entire pool area.

"You want anything?" Henry called out from the kitchen.

"No thanks."

He came out with an open bag of potato chips and set it on the table in front of her. "I found this on the floor in the front hall," he said, snapping a business card down on the table. "You know anything about it?"

Deirdre picked up the card. It had Joelen's name on it. "She was here this morning."

"I didn't know you two kept in touch."

"We didn't. Did you?"

"Why would I?" Henry said. "I barely remember her."

Deirdre let it go, but she knew that Henry remembered Joelen Nichol. Remembered her well, and not just because she'd made headlines. Henry, who'd never wanted Deirdre within fifty feet of him and his friends, used to hang out with her whenever Joelen came over. Once she'd discovered the pair of them making out on the musty sleeper sofa that her parents stored in the garage.

Sometimes, on nights when she was sleeping over at Joelen's, Henry would show up late and toss pebbles at Joelen's bedroom window so she'd come down. One night Henry's pebble missed Joelen's window and hit Bunny's instead. Bunny's boyfriend, Antonio Acevedo—the man everyone called "Tito"—had come whaling out of the house, armed with a baseball bat. Lucky Henry had ridden over on his bicycle and could get the hell out of there before he got hurt.

"She wasn't here to see me," Deirdre said. "She was here to talk to Dad."

Henry's look darkened. "Why'd she want to talk to Dad?"

Deirdre pointed to the setting sun logo and SUNSET REALTY above Joelen's name. "Just guessing. She's a Realtor. He's selling a house."

"I thought he already talked to a Realtor."

Deirdre shrugged. "All I know is she was here. She said she had a meeting with him. She freaked out when the police arrived."

"I'll bet she did."

"Don't be mean. I remember, you liked her."

"Sure I liked her," Henry said. "We had fun. Fooled around. But it was never serious. I haven't talked to her since she killed—"

"Supposedly killed."

Henry stood at the glass door and looked out into the yard. "Hey, she confessed."

CHAPTER 5

The story had made national news—DAUGHTER KILLS STAR'S BOYFRIEND.

It had happened on a night when Deirdre was sleeping over at the Nichols' house, late after one of Bunny Nichol's lavish parties. Bunny's boyfriend, Antonio "Tito" Acevedo, was stabbed to death in her bedroom.

Deirdre didn't find out about the murder for days after because she was in the hospital. Her father—he and Gloria had been among the guests at the party earlier—had come back in the middle of the night to take her home. He'd carried her, half-asleep, out to his car. On the way home, his car skidded off the road and she was thrown out.

She'd spent weeks in Northridge Hospital—Arthur had insisted the ambulance take her there because of their excellent reputation rehabbing Vietnam vets. After multiple operations, skin grafts, and physical therapy, the doctors finally conceded that the damage to her femoral nerve was permanent. She'd never be able to move her hip or bend and straighten her leg. She'd never feel heat, or cold, or pain, or even a gentle touch on the front of her thigh. Over time, the muscles would atrophy.

No one had warned her how much she'd come to cherish what she'd once been—unremarkable and nearly invisible. Instead, her mere presence would attract uneasy stares.

Desperate for anything to distract her from the pain and uncertainty of her ordeal, Deirdre had found a newspaper someone had left in the hospital visitors' lounge and read about the murder. After that she watched the nightly news, first from her hospital bed and later from the living room couch, as the story of the murder, photographs of the crime scene, and the lives of Bunny and Joelen Nichol and Tito Acevedo were endlessly dissected and fed to an audience ravenous for every sordid detail. Later, when Deirdre was strong enough to visit the public library, she surreptitiously tore news articles from the public copies of the *L.A. Times* and stole away with them so she could read and reread their accounts of the murder and inquest that followed.

The cause of death was a single knife thrust to the solar plexus; apparently Tito had dropped like a stone. "I did it," Joelen had told the police, who must have arrived at the house after Arthur drove off with Deirdre.

At the hearing, the coroner made a big deal about the lack of defensive wounds. Why hadn't he tried to protect himself? But that didn't seem at all far-fetched to Deirdre. Tito Acevedo, who carried a roll of hundred-dollar bills and a silver monogrammed gun-shaped Zippo lighter in his trouser pocket, would never have seen it coming. He wouldn't have been the slightest bit afraid when Joelen came at him, all of fifteen years old, a hundred pounds, dressed in that flowered cotton granny gown she wore whenever Deirdre slept over.

"He ran into my knife," Joelen told the coroner's jury.

That ten-inch kitchen knife was scrutinized, as was the night-gown Joelen had been wearing. An expert who testified was skeptical. Why wasn't there more blood? he wanted to know. From the wound Tito suffered, there should have been more.

But far more compelling than the presence or absence of blood evidence or defensive wounds was the dramatic testimony of Joelen's tearful movie star mother. Bunny Nichol sat in the witness box wearing a dark suit and a blouse with a ruffled collar that swathed her neck like a bandage. Her jet-black hair was pulled back in a severe French twist. In the black-and-white television images, there were bruises under her eye and over her jaw, livid against skin that was otherwise flawless as bone china. She answered each question posed to her in a calm, quiet voice. It had been odd to see Sy Sterling, whom Deirdre had known forever as her father's best friend, performing his courtroom role on TV, a scaled-down Perry Mason.

"Why did you stay with a man who beat you?" Sy had asked, just a trace of his Russian accent surfacing: *bitt you.*

"I was afraid," Bunny said, staring down and kneading her hands together. "I had to do anything and everything he wanted or he said he'd ruin my face. He said I'd be sorry if I ever tried to leave him. He said if I told anyone, he'd get me where it hurt most. I knew what he meant." She'd paused and her audience, including Deirdre, had leaned into the silence. "My daughter. He would have killed us both."

Deirdre had heard Bunny and Tito fighting some nights when she'd slept over. Angry shouting matches. Breaking glass. She could easily imagine herself in Joelen's place, listening to Tito's escalating threats and growing more and more terrified. Formulating a plan. Creeping downstairs to the kitchen. Pulling open a drawer and selecting the longest, sharpest, pointiest knife she could find. As she climbed the stairs, had Joelen thought about what would happen after? Did she hesitate as she approached the closed bedroom door? Did she have second thoughts as she stood in the hallway, screwing up her courage? Something must have spurred her to act at the moment that she did. Maybe it had been the sound of furniture breaking. Or a fist slammed into a wall. Or Bunny crying out.

It hadn't taken the coroner's jury long. After a few hours they ruled. Justifiable homicide. It wasn't *not guilty*, but it wasn't *guilty*, either. The verdict kept Joelen from being indicted for murder.

A real "David slays Goliath tale" was the verdict rendered by the TV newscaster Deirdre watched, lying on the living room sofa recovering from her first operation. She tried to call Joelen after the hearing but no one answered. She wrote to her but got no response. She begged her parents to drive her over there but they said there was no point to that. Bunny had left town. It was as if Deirdre's friend had vanished into thin air.

For months after, Bunny Nichol kept an uncharacteristically low profile too. Then came the news that she was back in town and married to a handsome young TV soap opera star, Derek Hutchinson. A few months later, the papers ran a photograph of the happy couple with a baby. Reporters were a tad more discreet in those days: Deirdre didn't remember the press commenting on the obvious fact that Bunny Nichol had been pregnant when she'd had her final fight with Antonio Acevedo. Pregnant when she testified on nationwide TV. No one was surprised that the baby boy, with his head of dark hair, olive skin, and dark eyes, resembled Antonio Acevedo a whole lot more than he resembled Derek Hutchinson, who was slender and fair. But those rumors were a gentle breeze compared to the shit storm that got kicked up a few years later when Derek Hutchinson died of AIDS, one of the sad first wave that took out so many of Hollywood's most talented.

Deirdre was finally well enough to return to school near the end of the academic year. At least she walked back into class on crutches, not in a wheelchair. Even the high and mighty Marianne Wasserman was friendly and solicitous, organizing a posse of her friends to carry

Deirdre's books between classes. It made Deirdre queasy now, remembering the small amount of celebrity status she'd found herself basking in simply because she'd been Joelen's friend. Even as she'd traded on her friendship with Joelen, it had occurred to her how toxic notoriety could be.

ABOUT THE AUTHOR

HALLIE EPHRON is the bestselling, award-winning author of suspense novels. Her novels have been finalists for the Edgar, Anthony, and Mary Higgins Clark awards. With *Night Night, Sleep Tight*, she takes readers back to early-'60s Beverly Hills, a time and place she knows intimately. She grew up there, the third of four daughters of Hollywood screenwriting duo Henry and Phoebe Ephron, contract writers for Twentieth Century Fox who wrote screenplays for classics like *Carousel* and *Desk Set*. Ms. Ephron's novels have been called "Hitchcockian" by *USA Today*, and "deliciously creepy" by *Publishers Weekly*. Her award-winning bestseller *Never Tell a Lie* was made into a movie for the Lifetime Movie Network. Her essays have been broadcast on NPR and appeared in magazines including *More*, *Writer's Digest*, and *O: The Oprah Magazine* ("Growing Up Ephron"). She writes a regular crime fiction book review column for the *Boston Globe*. Ms. Ephron lives near Boston with her husband.